DISTRIBUTED REMOTE SENSING FOR NAVAL UNDERSEA WARFARE

ABBREVIATED VERSION

Committee on Distributed Remote Sensing for Naval Undersea Warfare
Naval Studies Board
Division on Engineering and Physical Sciences

NATIONAL RESEARCH COUNCIL
OF THE NATIONAL ACADEMIES

THE NATIONAL ACADEMIES PRESS
Washington, D.C.
www.nap.edu

THE NATIONAL ACADEMIES PRESS 500 Fifth Street, N.W. Washington, DC 20001

NOTICE: The project that is the subject of this report was approved by the Governing Board of the National Research Council, whose members are drawn from the councils of the National Academy of Sciences, the National Academy of Engineering, and the Institute of Medicine. The members of the committee responsible for the report were chosen for their special competences and with regard for appropriate balance.

This study was supported by Contract No. N00014-05-G-0288, DO #1 between the National Academy of Sciences and the Department of the Navy. Any opinions, findings, conclusions, or recommendations expressed in this publication are those of the author(s) and do not necessarily reflect the views of the organizations or agencies that provided support for the project.

International Standard Book Number-13: 978-0-309-10180-6
International Standard Book Number-10: 0-309-10180-8

Additional copies of this report are available from:

Naval Studies Board, National Research Council, The Keck Center of the National Academies, 500 Fifth Street, N.W., Room WS904, Washington, DC 20001; and

The National Academies Press, 500 Fifth Street, N.W., Lockbox 285, Washington, DC 20055; (800) 624-6242 or (202) 334-3313 (in the Washington metropolitan area); Internet, http://www.nap.edu.

Printed in the United States of America

THE NATIONAL ACADEMIES

Advisers to the Nation on Science, Engineering, and Medicine

The **National Academy of Sciences** is a private, nonprofit, self-perpetuating society of distinguished scholars engaged in scientific and engineering research, dedicated to the furtherance of science and technology and to their use for the general welfare. Upon the authority of the charter granted to it by the Congress in 1863, the Academy has a mandate that requires it to advise the federal government on scientific and technical matters. Dr. Ralph J. Cicerone is president of the National Academy of Sciences.

The **National Academy of Engineering** was established in 1964, under the charter of the National Academy of Sciences, as a parallel organization of outstanding engineers. It is autonomous in its administration and in the selection of its members, sharing with the National Academy of Sciences the responsibility for advising the federal government. The National Academy of Engineering also sponsors engineering programs aimed at meeting national needs, encourages education and research, and recognizes the superior achievements of engineers. Dr. Wm. A. Wulf is president of the National Academy of Engineering.

The **Institute of Medicine** was established in 1970 by the National Academy of Sciences to secure the services of eminent members of appropriate professions in the examination of policy matters pertaining to the health of the public. The Institute acts under the responsibility given to the National Academy of Sciences by its congressional charter to be an adviser to the federal government and, upon its own initiative, to identify issues of medical care, research, and education. Dr. Harvey V. Fineberg is president of the Institute of Medicine.

The **National Research Council** was organized by the National Academy of Sciences in 1916 to associate the broad community of science and technology with the Academy's purposes of furthering knowledge and advising the federal government. Functioning in accordance with general policies determined by the Academy, the Council has become the principal operating agency of both the National Academy of Sciences and the National Academy of Engineering in providing services to the government, the public, and the scientific and engineering communities. The Council is administered jointly by both Academies and the Institute of Medicine. Dr. Ralph J. Cicerone and Dr. Wm. A. Wulf are chair and vice chair, respectively, of the National Research Council.

www.national-academies.org

COMMITTEE ON DISTRIBUTED REMOTE SENSING FOR NAVAL UNDERSEA WARFARE

ARTHUR B. BAGGEROER, Massachusetts Institute of Technology, *Co-chair*
BRIG "CHIP" ELLIOTT, BBN Technologies, *Co-chair*
JAMES G. BELLINGHAM, Monterey Bay Aquarium Research Institute
E. ANN BERMAN, Tri-Space, Incorporated
D. RICHARD BLIDBERG, Autonomous Undersea Systems Institute
DANIEL R. BOWLER, Lockheed Martin Corporation
DAVID L. BRADLEY, Applied Research Laboratory, Pennsylvania State University
ALBERT H. KONETZNI, JR., Martinez, Georgia
WILLIAM A. LaPLANTE, Applied Physics Laboratory, Johns Hopkins University
TERRY P. LEWIS, Raytheon Company
THOMAS V. McNAMARA, Charles Stark Draper Laboratory
L. DAVID MONTAGUE, Menlo Park, California
DOUGLAS R. MOOK, The Aptec Group
JOHN E. RHODES, Balboa, California
JAMES WARD, Lincoln Laboratory, Massachusetts Institute of Technology
DANA R. YOERGER, Woods Hole Oceanographic Institution

Staff

CHARLES F. DRAPER, Director
ARUL MOZHI, Study Director
IAN M. CAMERON, Associate Program Officer (through May 21, 2007)
SUSAN G. CAMPBELL, Administrative Coordinator
MARY G. GORDON, Information Officer
AYANNA N. VEST, Senior Program Assistant (through June 9, 2006)
SIDNEY G. REED, JR., Consultant
RAYMOND S. WIDMAYER, Consultant

NAVAL STUDIES BOARD

Staff

CHARLES F. DRAPER, Director
ARUL MOZHI, Senior Program Officer
EUGENE J. CHOI, Program Officer (through May 18, 2007)
IAN M. CAMERON, Associate Program Officer (through May 21, 2007)
SUSAN G. CAMPBELL, Administrative Coordinator
MARY G. GORDON, Information Officer
AYANNA N. VEST, Senior Program Assistant (through June 9, 2006)

Preface

This study responds to the request from the former Chief of Naval Operations (CNO) that the National Research Council's (NRC's) Naval Studies Board conduct an assessment of distributed remote sensing (DRS) for naval undersea warfare.[1] This request occurs at a time when the overarching guidance from *Naval Power 21* articulates the need for the Department of the Navy to ensure access worldwide for military operations.[2] Quiet diesel electric submarines and increasingly sophisticated mines available to potential enemies are a threat to such access, especially for missions involving port ingress and egress, rapid transit through choke points, and operations in deep as well as shallow littoral waters.

The Department of the Navy has reorganized to give undersea warfare (USW) greater prominence and has conducted several experiments to explore the potential of new DRS approaches to improve its capability to counter the antisubmarine warfare (ASW) threat.[3] The *CNO Guidance for 2006* includes the objective of ensuring the ability to detect and hold at risk adversary submarines and of shortening the detection-to-engagement time line in both deep and shallow waters. The CNO has been encouraging the rapid prototyping of enabling

[1]ADM Vern Clark, USN, CNO, letter dated April 18, 2005, to Dr. Bruce M. Alberts, President, National Academy of Sciences.

[2]Hon. Gordon England, Secretary of the Navy; ADM Vern Clark, USN, Chief of Naval Operations; and Gen James L. Jones, USMC, Commandant of the Marine Corps. 2002. *Naval Power 21*, Department of the Navy, Washington, D.C.

[3]See John R. Benedict. 2005. "The Unraveling and Revitalization of U.S. Navy Antisubmarine Warfare," *Naval War College Review*, Spring, pp. 93-120.

technologies to accomplish that goal, and *CNO Guidance for 2006* emphasizes the key role of improved sensors in meeting that objective.[4]

For the mine threat, which may require avoidance, rapid detection, and clearance of large numbers of mines, the use of DRS systems for persistent surveillance and cooperative, multiple, mobile sensors is needed to define areas of safe maneuver and safe passage rapidly, as described in the *FY07 U.S. Naval Mine Countermeasures (MCM) Plan*.[5]

The *Naval Transformation Roadmap*, which guides these efforts, states that transformational efforts in ASW are "focused on and planned for developing new operational concepts that leverage advanced technologies to improve broad-area surveillance, detection, localization, tracking, and attack capabilities against quiet adversary submarines,"[6] and that tranformational countermine warfare efforts will "employ sophisticated, networked unmanned surface, air, and underwater vehicles equipped with advanced technology systems . . . to quickly avoid or enter and safely clear dangerous mined areas."[7]

Previous reports of the Naval Studies Board, such as *Mine Countermeasures Technology* (U); *Technology for the United States Navy and Marine Corps, 2000-2035, Volume 7: Undersea Warfare;* and *Naval Mine Warfare: Operational and Technical Challenges for Naval Forces,*[8] have recommended that high priority be accorded the development of networked, distributed sensor fields, including unmanned platforms for submarine detection and mine countermeasures (MCM), or, more appropriately, countermine warfare.[9]

The essential features of a DRS system for USW include the following: a sensor field involving a number of fixed and/or moving nodes to conduct sur-

---

[4]Chief of Naval Operations (ADM Michael Mullen, USN). 2005. *CNO Guidance for 2006: Meeting the Challenge of a New Era,* Department of the Navy, Washington, D.C., October 30, p. 5.

[5]Department of the Navy. 2006. *FY07 U.S. Naval Mine Countermeasures (MCM) Plan: Vision, Roadmap, and Program,* Washington, D.C., February.

[6]Hon. Gordon England, Secretary of the Navy; ADM Vern Clark, USN, Chief of Naval Operations; and Gen James L. Jones, USMC, Commandant of the Marine Corps. 2002. *Naval Transformation Roadmap, Power and Access . . . From the Sea,* Department of the Navy, Washington, D.C., February 6, p. 19.

[7]Hon. Gordon England, Secretary of the Navy; ADM Vern Clark, USN, Chief of Naval Operations; and Gen James L. Jones, USMC, Commandant of the Marine Corps. 2002. *Naval Transformation Roadmap, Power and Access . . . From the Sea,* Department of the Navy, Washington, D.C., February 6, p. 21.

[8]Naval Studies Board, National Research Council, 1993, *Mine Countermeasures Technology* (U), National Academy Press, Washington, D.C. (classified); Naval Studies Board, National Research Council, 1997, *Technology for the United States Navy and Marine Corps, 2000-2035: Becoming a 21st-Century Force, Volume 7: Undersea Warfare,* National Academy Press, Washington, D.C.; Naval Studies Board, National Research Council, 2001, *Naval Mine Warfare: Operational and Technical Challenges for Naval Forces,* National Academy Press, Washington, D.C.

[9]The term "countermine warfare" includes not only local measures to detect and clear mines, but also the intelligence and other support activities important to countering the threat of mines at sea and in the approaches to shore.

veillance, detection, and localization of submarines or mines; communications links to transmit data from the sensor subsystem to a processing facility or unit; and a communications center to receive results from the processing facility or unit, to combine them with other intelligence for intelligence, surveillance, and reconnaissance (ISR), and in a time of hostilities to cue available attack assets to locations where targets can be found more precisely and attacked or neutralized. On the battlefield, all of these systems must carry out their functions for DRS to be effective. DRS systems can be deployed before or during hostilities and can be fixed, drifting, or propelled. There are now several examples of sensor nodes, but not systems, that have the attributes listed above.

No single type of sensor alone is adequate to the task, although under the surface of the water acoustic sensors dominate the tactical sensors. Acoustic sensors are divided into passive and active categories. All other sensors are included in the category nonacoustic sensors. Multiple independent detections of a target are very effective at providing a high probability of detection at a low false-alarm rate. For that reason and because various sensors have complementary capabilities, multisensor fields can offer advantages in demanding situations, although their use may limit detection ranges, since some sensors excel at greater detection ranges but with less precision. The use of more sensors can help only if each is treated within its performance envelope.

Historically, the Navy has operated a number of distributed remote sensing systems for ASW. Air-dropped sonobuoys have become increasingly sophisticated since their first use in World War II. At first passive, sonobuoy systems can now involve active sources. During the Cold War, the Navy operated the long-range Sound Surveillance System—a passive DRS system.[10] Passive acoustic sensor arrays are towed by Surveillance Towed Array Sensor System (SURTASS) ships and submarines, with increased capability for signal processing and networking, specifically in deeper water.

When the Soviets quieted their submarines in the 1980s, the response of the U.S. Navy was to develop short-range passive, acoustic arrays for deployment on the ocean bottom in critical operational areas. The Fixed Distributed System (FDS) connects a number of distributed sensor clusters by cable with one another and to a signal-processing and communications center on land. The FDS's location is necessarily limited to areas and times in which it is safe to deploy. The follow-on Advanced Deployable System (ADS) has similar technology, is designed to be deployable by ship or submarine in deep or shallow water, and connects (by cable) to a processing and communications buoy on the surface. DRS systems can include nonacoustic sensors able to detect submarine-associated emissions or phenomena.

The question addressed in the present study is this: Can the U.S. Navy use its

[10]All DRS ASW systems have not been acoustic. During World War II, the 10th Fleet exploited high-frequency electromagnetic emission from German submarines, intercepted by distributed fixed U.S. and British stations, to help win the Second Battle of the Atlantic.

advanced distributed remote sensing technology effectively to counter the threats it faces in carrying out its mission? At issue are the technical feasibility, concept of operations, networked-systems definition, processing requirements, cost, and effectiveness measures for DRS in support of specific undersea warfare missions. While all of these are very challenging technically, the deliberations of the NRC's Committee on Distributed Remote Sensing for Naval Undersea Warfare led to a positive answer; nevertheless, significant research and development (R&D) as well as training need to be carried out for advanced DRS technology to be used routinely by fleet operators.

TERMS OF REFERENCE

At the request of the former Chief of Naval Operations,[11] the Naval Studies Board of the National Research Council conducted an assessment of distributed remote sensing for naval undersea warfare. Specifically, the study assessed the following topic areas:

- Undersea (antisubmarine and countermine) warfare missions in future operating areas, including for port access and egress, transit through choke points, and in shallow/sloping harsh acoustic littoral environments.
- Range of current approaches of DRS, and their utility in possible concepts of operation for accomplishing specific undersea warfare missions.
- Status and anticipated improvements in underwater sensors and communications for DRS and related platforms for specific undersea warfare missions.
- DRS requirements (sensor numbers, concept of employment, characteristics, target signature exploitation, aperture, connectivity, processing, longevity, latency of reporting, false alarm control measures, covertness/counterdetection, survivability, and reliability) to provide capabilities needed for specific undersea warfare missions.
- Technology for fusion capabilities of information from the underwater and above water sensors, and relationship to network-centric operations (FORCEnet, Global Information Grid).
- Underwater sensors and balance of vehicles to fixed nodes, including technology limitations affecting use of multiple unmanned undersea vehicles.
- Critical technologies, performance measures, and scaling relationships associated with DRS networks for specific undersea warfare missions.
- Expectations for improvements in applicable network technology and wireless connectivity, including information to engaged systems and decision making.
- Other operational factors such as ability for covert and rapid deployment, environmental sensitivity (including biological effects for active sonar systems), system, network and information security, training, and end-to-end logistics.

[11]ADM Vern Clark, USN, CNO, letter dated April 18, 2005, to Dr. Bruce M. Alberts, President, National Academy of Sciences.

THE COMMITTEE'S APPROACH

This report aims to provide a clear, near-term path by which useful distributed remote sensing systems can be applied rapidly to pressing naval undersea warfare problems and by which ongoing science and technology efforts can be channeled toward the most useful ends.

The approach of the committee was to create a tentative priority ranking of DRS applications by listing the key missions and matching these missions to the art of the possible. What is possible is a balance of what exists, what has been tried, and what can be operationally deployed in the near term taking into account constraints such as cost, technology maturity, and rules of engagement. The committee's hypotheses were then tested with strawman concepts and scenarios.

In its approach, the committee considered that DRS systems can be of great near-term use for naval missions, particularly those outlined below:

- Intelligence, surveillance, and reconnaissance in hostile littorals;
- Barriers for detection and cueing;
- Support for sea base protection; and
- Scouting and surveillance of high-risk transit routes.

During its assessment of DRS systems for naval undersea warfare, the committee focused on the ASW problem because that threat is growing rapidly and is new as far as the quiet, diesel electric submarines are concerned. Since two prior studies have addressed the countermine warfare topic,[12] the committee did not devote extensive attention to this issue. Instead, it concentrated on (1) reviewing the status of the Navy MCM program and comparing it with the recommendations in the two previous studies, and (2) considering briefly how the resulting planned organic MCM systems could be harnessed to compose DRS systems.

The committee (biographies of its members are provided in Appendix A) first convened in June 2005 and held additional meetings over a period of 7 months, both to gather input from the relevant communities and to discuss the committee's findings and recommendations.[13] Summary agendas for these meetings are provided in Appendix B.

The months between the committee's last meeting and the publication of the report were spent preparing the draft manuscript, gathering additional information, reviewing and responding to the external review comments, editing the report, and conducting the required security/public release review necessary to produce this version of the report that does not disclose information as described

[12]Naval Studies Board, National Research Council, 2001, *Naval Mine Warfare: Operational and Technical Challenges for Naval Forces,* National Academy Press, Washington, D.C.; Commander, U.S. Fleet Forces Command, 2002, *Mine Warfare—The Way Ahead: Long Range Planning and Concepts,* Norfolk, Va., February 5.

[13]During the entire course of its study, the committee held meetings in which it received (and discussed) materials that are exempt from release under 5 U.S.C. 552(b).

in 5 U.S.C. 552(b). It was mutually determined by the Department of the Navy and the National Research Council that the full report contained information as described in 5 U.S.C. 552(b) and therefore could not be released to the public in its entirety.

Acknowledgment of Reviewers

National Research Council (NRC) reports are reviewed in draft form by individuals chosen for their diverse perspectives and technical expertise, in accordance with procedures approved by the NRC's Report Review Committee. The purpose of this independent review is to provide candid and critical comments that will assist the institution in making its published reports as sound as possible and to ensure that the reports meet institutional standards for objectivity, evidence, and responsiveness to the study charge. The review comments and draft manuscripts remain confidential to protect the integrity of the deliberative process. Although the reviewers provide many constructive comments and suggestions, they are not asked to endorse the conclusions or recommendations nor do they see the final draft of reports before release. We wish to thank the following individuals for their review of the draft report:

Curtis G. Callan, Jr., Joseph Henry Laboratories, Princeton University,
Nicholas P. Chotiros, Applied Research Laboratory, University of Texas at Austin,
Henry Cox, Lockheed Martin Corporation,
Douglas J. Katz, VADM, USN (retired), Annapolis, Maryland,
John G. Schuster, Applied Physics Laboratory, Johns Hopkins University,
Robert C. Spindel, University of Washington,
Lawrence D. Stone, Metron, Inc., and
William A. Whitlow, MajGen, USMC (retired), Titan Corporation.

The review of the draft report was overseen by Robert J. Hermann, Global Technology Partners, LLC. Appointed by the NRC, he was responsible for making certain that an independent examination of the draft report was carried out in accordance with institutional procedures and that all review comments were carefully considered. Responsibility for the final content of NRC reports rests entirely with the authoring committee and the institution.

Contents

Executive Summary

INTRODUCTION

Since the early 20th century, the United States has been a nation with global responsibility for maritime commerce and power projection. Today the United States is recognized as the sole superpower, with its global reach often dependent on the capability of the U.S. Navy to put ships with bulk capacity offshore of virtually any country. This global reach is threatened, however, by the widespread acquisition of quiet, diesel electric submarines and inexpensive mines that can be very effective weapons in an adversary's littoral waters, whether shallow or deep.

The question then becomes: Can the United States, by implementing distributed remote sensing (DRS) technology, counter this asymmetric threat? This study, which was requested by the former Chief of Naval Operations, addresses the state of the art of and the challenges for DRS and related technologies, assesses current shortfalls, and makes recommendations toward rapidly implementing DRS systems to counter the growing submarine threat as well as problems of countermine warfare.[1] Most concepts for DRS are not mature, and there are many challenges in integrating them for use as a military system; neverthe-

[1]The essential features of a DRS system for undersea warfare include the following: a sensor field involving a number of fixed and/or moving nodes to conduct surveillance, detection, and localization of submarines or mines; communications links to transmit data from the sensor subsystem to a processing facility or unit; and a communications center to receive results from the processing facility or unit, to combine them with other intelligence for intelligence, surveillance, and reconnaissance (ISR) and, in a time of hostilities, to cue available attack assets to locations where targets can be found more precisely and attacked or neutralized. On the battlefield, all of these systems must carry out their functions for DRS to be effective. Before hostilities, DRS can provide ISR for analysis and archiving.

less, the National Research Council's (NRC's) Committee on Distributed Remote Sensing for Naval Undersea Warfare believes that, as described in this report, there are some selective short-term and many long-term opportunities to develop useful naval systems. The committee emphasizes that much remains to be done—especially for those systems involving significant automation and networking.

CHALLENGES OF DISTRIBUTED REMOTE SENSING FOR ANTISUBMARINE WARFARE

One of the first observations that became evident to the committee was that DRS has almost as many different interpretations as government program offices supporting efforts in the area. This lack of consensus makes it very difficult to focus on definitive goals. The list below suggests the most important operational components needed to parse a DRS system for the prosecution of antisubmarine warfare (ASW).

- *Cueing.* Cueing is the process of achieving sufficiently accurate localization on a target that a commander can successfully carry out a follow-up prosecution of the target. Typically, prosecution, once cued, involves the redetection and refinement of localization. A subtle distinction needs to be made with respect to types of cueing operations: *Wide-area search*, typically rated in terms of a "sweep rate," can be used dynamically; in contrast, *area coverage* such as that provided by the Fixed Distributed System (FDS) or the Surveillance Towed Array Sensor System (SURTASS) is used over a (relatively) fixed region. The committee did not come across a DRS concept that could change the current methods of cueing and lead to a more speedy and effective prosecution of targets, short of a complete blanket deployment of a large number of fixed arrays.
- *Detection.* At face value, the concept of detection by means of a distributed system has many advantages. The diversity of opportunity for detection by widely spaced acoustic sensors seems attractive. Consequently one can ask: How can the observations made from many distributed sensors, each with small gain, be related to that from one large sensor with high-array gains?[2] Work has been done recently by Task Force Antisubmarine Warfare on detections by many closely spaced, small, short-range acoustic and combined acoustic and nonacoustic sensors (e.g., Deployable Autonomous Detection System), but these approaches to date have not been successful when placed in realistic environments. Certainly one of the goals of a DRS program should be to gain an understanding of distributed detection by passive and active, acoustic and nonacoustic sensors, both theoretically and practically.

[2]Passive detection is fundamentally limited by signature level, noise level, and achievable array and processing gains.

• *Classification and automation.* Classification remains the most difficult aspect of passive acoustic ASW, especially in a very noisy and cluttered region such as the western Pacific. Classification is now best done passively by highly skilled acoustic-intelligence (ACINT) operators who have nearly 10 years of training before being ranked in this specialty. A computer algorithm has not yet been found that can mimic a skilled ACINT operator, although more-routine classifications can be off-loaded to computer algorithms. The distributed system should help by providing multiple perspectives and reinforcing features, but many of the key issues of distributed classification and the tactics to exploit such a system remain unresolved and, in some cases, unexplored.

• *Communications.* Without communications, no sensor system can operate as a distributed system. Radio frequency (RF), cables, and acoustic communications (ACOMMS) are possible means. One direct undersea approach would use ACOMMS. While such systems are available commercially and can reasonably cover a 35-mile range, they are hardly covert. Minimizing the probability of interception by pointing beams and spread-spectrum coding is possible, but energy detection is still quite likely. Larger-power and lower carrier frequencies can go beyond the 35-mile range, but, again, these frequencies aggravate the covertness problem. With ACOMMS there is also the problem of spectral management, because the available bandwidth is so limited. Routing through a network to ensure reliable coupling among the nodes is also possible. The same issues apply to RF communications, and cables need to be deployed.

• *Distributed network control.* In using mobile unmanned undersea vehicles (UUVs) or gliders, there may be the possibility of closing the range to the target, if there is a forward interception point, to improve the signal-to-noise ratio (SNR) and exploit lower-level features for classification. Without a low probability of false alarm, however, the UUV or glider would as likely move away from the target as toward it. This implies the need for some form of metric for optimizing the sensor configuration, with significant uncertainties about individual performance measures. Again, good theoretical and practical support is not available at present, and fundamental research is needed.

• *Sensor distribution design versus speed mismatch.* Current UUVs run at 3 to 5 knots and can sprint briefly at higher speeds. This low speed is consistent with that of a high-end diesel electric submarine at patrol speed, but the latter requires very little additional speed to exceed that of any component of the network of UUVs. This speed mismatch leads to an issue of how to design clusters of distributed systems in such a way that a patrolling enemy submarine will enter a local network and then be detected and tracked with high probability. The same issue applies to the placement of fixed arrays. Previous work on the optimality of sonobuoy patterns is relevant, but, again, work is needed on better designs for distributed networks.

• *Rules of engagement.* How is it determined when to deploy a distributed system during prehostilities in a potential battlespace? Naturally this involves the complexity and cost of the system. Some operations—such as activating a barrier—could be considered hostile. How much tactical advantage is an adversary allowed to accrue before U.S. forces fire the first shot to prevent U.S. ships and sailors from being at unacceptable risk? Guidelines need to be developed for employing these systems, which in the aggregate will be expensive, but they certainly should be used if U.S. ships and sailors will be at risk.

- *Logistics and deployment.* Operating the field of sensors for a DRS system as well as deploying them, or putting them in place, requires considerable logistical support. For a fleet of UUVs for DRS, a large component of a ship, or perhaps several modules of a littoral combat ship (LCS), would be required to ferry, deploy, and reseed UUVs as needed, in addition to whatever role the ship might play as a gateway back to command authority. The UUV-ferrying operation would consume substantial volume, whether on deck or below; deployments must be simple and safe in high sea states; reseeding calls for spares and a crew to maintain the UUVs or arrays. None of these requirements is trivial.
- *Concept of operations (CONOPS).* One can postulate cueing a deployment in semi-real time or establish a more permanent deployment in a region of opportunity, but the most pressing issue for a DRS system is to develop a viable CONOPS. The challenge is for a DRS network to provide the area coverage, localization, and tracking capabilities that are militarily useful. Assuming that the classification can be resolved and the target put into localization and track, the network must reconfigure to maintain track and trail for as long as possible before passing the target vector onto the next DRS complex. The capabilities are hardly all in hand. Thus, the science and technology (S&T) and the research and development (R&D) needed to enable a DRS concept are extensive. Providing them will require significant funding and a focused effort by the Navy.

CONCLUSIONS, FINDINGS, AND RECOMMENDATIONS

The committee's key conclusions are as follows:

- Today's distributed remote sensing technology is already adequate for tackling some very specific urgent naval operational needs.
- The Navy's approach to employing DRS systems will need to focus on defining and concentrating on one or two pressing problems—not the entire design space.
- Candidate DRS solutions need to address the entire end-to-end (detection-to-prosecution) system and CONOPS for a given operational task, with appropriate system analysis, architecture, and external interfaces.
- The Navy needs to focus on the art of the possible and get systems into the field and then improve them—that is, try to fill some technology gaps as testing proceeds.

• Interoperability and open architectures[3] will be key to affordability and the quick fielding of new capabilities.

The committee's findings and recommendations are presented below.

Finding 1: DRS can provide significant, near-term capabilities for some specific needs such as intelligence, surveillance, and reconnaissance (including the acquisition of acoustic and environmental data) and for sensor barriers. While DRS is deemed critical to Navy operations in important regions as a force multiplier, cost saver, and means of keeping sailors out of harm's way, there is no dedicated program-level office responsible for the development and implementation of DRS.

The committee notes that DRS systems can be a force multiplier, but *it emphasizes very strongly* that most of the potential has not been realized, except for the current ASW capability of air-dropped systems such as sonobuoys, including directional command-activated sonobuoy systems and air-deployable active receiver extensions (e.g., the Extended Echo Ranging [EER] family of systems).

The committee's assessment is that the technological capabilities of remote systems have steadily improved—largely driven by micro- and nanoelectronics. Fixed arrays can have more sensors; also, greater in situ processing and memory as well as coupled cables with high-bandwidth fiber optics are available now compared with 5 years ago. UUVs, both propelled and gliders, now have longer mission durations and larger payloads. Clandestine data acquisition is an easier task for DRS than use as a barrier for detection, classification, localization, and tracking (DCLT) and requires minimum distributed-network control. Adaptive sampling algorithms could also be of use. Fixed or portable barriers are more complicated because of the current state of the art in target classification, especially if the barriers are near frequently traversed routes, but one could consider them simply as a distributed collection asset that provides an alert by some means. Depending on the needs and CONOPS, some selected missions in local areas could be accomplished in the short term.

Nevertheless, the committee notes the very apparent weakness or even lack of CONOPS for DRS systems, the lack of methodical analyses or systems engineering, the absence of operational and technical architectures, and the general lack of definition of how DRS might be used in practice—that is, how DRS usage maps into the various phases of conflict, logistics, and rules of engagement. In summary, the integration of the technological capabilities and experiments with the very pressing operational needs for which DRS seems to be well suited has been difficult. Like many similar complex warfighting problems, the issue is not

[3]Open architectures involve the use of widely accepted and available specifications, standards, products, and design practices to produce systems that are interoperable, easy to modify, and extensible. The committee notes that "open" does not imply being unclassified.

"operational pull" or "technology push" but the need for the establishment of an iterative dialogue between the two. Therefore, the goal ought to be to bring the technological and the operational communities together to develop CONOPS based on disciplined analysis and experiment. A well-defined vision supported by a program office should help this process significantly.

This problem urgently needs the attention of the Assistant Secretary of the Navy for Research, Development, and Acquisition (ASN[RDA]). The ASN(RDA) needs to assess the management of myriad DRS-related activities in research, development, experimentation, and testing to determine if a new program office is warranted or if a reorganization within an existing program executive office would provide the leadership, cohesiveness, and management to focus on providing near-term DRS capabilities to counter important threats. The near-term DRS capabilities should then be embodied in a prototype DRS system that would be tested, experimented with, and developed and assessed in a spiral and persistent manner with a strong linkage to S&T. Further, spiral development should be organized for a long-term commitment following the example of the Advanced Processor Build/Acoustic Rapid COTS (commercial off-the-shelf) Insertion (APB/ARCI) program.

Recommendation 1: In order to bring to fruition distributed remote sensing systems for significant, near-term capabilities, the committee recommends the following:

- The Deputy Chief of Naval Operations for Integration of Capabilities and Resources (N8) should ensure that all undersea DRS requirements and resources become resident within a single directorate in the Office of the Chief of Naval Operations, so as to maximize synergy in funding resources, R&D, and systems integration for the achievement of near-term results.
- The Commander, Fleet Forces Command (CFFC), and the Navy Warfare Development Command (NWDC) should determine the requirements for and the CONOPS and concepts of employment of the several passive and active acoustic and other DRS systems for antisubmarine warfare within the following two areas for near-term promise and deployment: (1) peacetime intelligence, surveillance, and reconnaissance; and (2) fixed and/or portable barriers during hostilities.
- The Assistant Secretary of the Navy for Research, Development, and Acquisition should establish an office chartered to implement these two capabilities. The Program Executive Office for Integrated Warfare Systems (PEO[IWS]) appears to be a logical location for this office.
- The PEO(IWS) (or the new office) should ensure that the CONOPS is developed under realistic conditions for both peacetime and hostilities. It should guide program management efforts and schedules and perform disciplined and detailed systems analysis—that is, war games with expected system performance (with less than ideal detection, classification, localization, and tracking as well

as without pre-staged force levels before the initiation of combat) and realistic rules of engagement from the national command. The goal should be to define and concentrate on the two pressing problems (as suggested above), not the whole design space.

- The efforts of the PEO(IWS) (or the new office) should lead to a spiral-development cycle modeled on the APB/ARCI program. Attributes of this model are as follows:
  - —Basic implementation hardware with more frequent software builds and complex hardware components dropped in during each spiral;
  - —A data-driven, robust peer review process to vet and test proposed improvements and algorithms on the basis of available field data;
  - —A disciplined and consistently resourced activity, strongly coupled to the science and technology program of the Office of Naval Research (ONR), to serve as input to future DRS systems that focus on key S&T enablers (see Recommendation 4); and
  - —Open architectures and well-defined interoperability standards.
- The PEO(IWS) should consider the capabilities of the Mission Reconfigurable UUV System (MRUUVS), or its equivalent, to be a suitable starting point for each of the peacetime and barrier DRS systems. Final equipment selection should be based on careful systems analysis.

For the long term, the goal is a suite of both fixed and mobile assets with appropriate in-node DCLT capability, covertness, bandwidths, persistence, and communications consistent with their location, for example, with or without defense, according to the CONOPS established as recommended above.

Finding 2: The Navy's capability for broad-area search of sufficient accuracy for cueing for antisubmarine warfare has been badly eroded with respect to diesel electric submarines, and especially the newer high-end types. With the exception of the Fixed Distributed System, the committee found no existing capability for broad-area search and cueing. There is an urgent need for such systems in selected deep- and shallow-water regions. The problem is extremely difficult, but the committee believes that the best (available) answer is similar to the objectives that have been stated for the Advanced Deployable System (ADS), with both passive and command-active capability and with long duration and fixed distribution.[4]

The committee believes that long-duration, fixed distributed assets are key to achieving and maintaining battlespace control over broad areas. Situational awareness comes with a long-term investment in acquiring knowledge of a region,

[4]The committee does not challenge the objectives of the ADS, but rather the CONOPS for deployment, its ocean engineering, and the use of an LCS for a gateway.

including its oceanographic, shipping, and fishing environments. In addition, the committee believes that for specific regions rather than broad areas, a rapidly deployable, combined active and passive system for ASW area search in both shallow and deep water is essential for countering important threats, such as the modern diesel electric submarine. Moreover, such a system is technically and operationally feasible.[5] The kinematics and physics of the problem strongly suggest that the only practical way to achieve this capability is with a deployable, multinode passive-plus-command-active system. The Navy's current path, however, will not provide such capabilities in the foreseeable future.

The committee's concerns with the implementation of the shallow-water ADS program were significant and are addressed in Finding 3, below. The other various distributed system efforts for both deep and shallow water now being prototyped and demonstrated in the Navy S&T program or Task Force Antisubmarine Warfare have shown some promising aspects; however, there is a significant lack of awareness and symbiotic efforts among these programs. Alternatively stated, the Navy's focus is not cohesive. These programs now lack clear transition paths and have had very limited supporting analysis and system design work for their use as an ASW system.

Recommendation 2: The ASN(RDA) and the Deputy Chief of Naval Operations for Integration of Capabilities and Resources (N8) should commission a design study that will perform a robust analysis of the systems concepts as well as the CONOPS consistent with Recommendation 1, including both acoustic and nonacoustic sensing, for an engineering effort to develop a combined active and passive, rapidly deployable distributed remote system for antisubmarine warfare. Such a system should effectively replicate the expected passive performance of the Advanced Deployable System, but it should be augmented with an active capability. Attributes of such a design study should include the following:

- Well-staffed, high-profile, and quick turnaround of 6-12 months duration;
- Comprehensive and quantitative prediction and optimization of the design trade space for important regions of interest, including desired probability of detection/probability of false alarm, area coverage, persistence, and costs;
- The identification of S&T key enablers and technical barriers;
- The development of a common architecture in accordance with the Department of Defense Architecture Framework (DODAF)—for example, operational and systems architectures as part of a future deliverable; and
- Conceptual designs with modeled performance.

[5]A DRS system for littoral and shallow water will likely be different from a DRS system for deep water in several ways.

The DASN(IWS) should support the PEO(IWS) to execute this passive and active distributed remote system modeled on the APB/ARCI program referred to in Recommendation 1. Note that this does not imply that there needs to be a single system, but rather a suite of systems that covers the Navy's needs in areas of interest. The goal should be to focus initially on the art of the possible and then to make improvements.

Finding 3: The Advanced Deployable System program was intended to provide the rapidly deployable regional search capability discussed in Finding 2. The committee is concerned about the current implementation, CONOPS, and execution of the ADS program.[6]

The ADS program was begun in the early 1990s. The underlying concept of ADS has been shown to work against high-end threats. However, the system has undergone no less than three major design changes—some because of technological advances and some because of decisions to change the delivery platform from submarine to ship. An original requirement for air delivery was abandoned for funding reasons. These multiple redirections of the ADS program have affected the development of the CONOPS and concept of employment and, of course, the budget and schedule. The history of requirements changes and creep, delivery-platform changes, oversight management changes, and inconsistent funding levels, and the lack of consensus and coordination among the requirements, acquisition, and operational communities—involving both delivery-platform and program developers—have tainted a concept and program so valuable and needed.

Recommendation 3: The ASN(RDA) and the Deputy Chief of Naval Operations for Integration of Capabilities and Resources (N8) should conduct a zero baseline review of the Advanced Deployable System program and requirements for a deployable system suitable for specific regions. This review should include at least the following:

- An assessment, led by the Director of Naval Intelligence, of the counter-detection of such a system from the adversary's perspective;
- An N8 assessment of re-establishing the submarine—for example, a nuclear-powered submarine (SSN) or a nuclear-powered, guided-missile submarine (SSGN)—as the delivery platform;
- An assessment by PEO(IWS) of the feasibility of delivery by a "heavy lift" unmanned undersea vehicle such as the large-diameter UUV Seahorse or by other heavy-lift air assets including those of the U.S. Air Force;

[6]As of this writing, the Navy plans major changes to the ADS program.

• An assessment by the Deputy Assistant Secretary of the Navy (Ships) (DASN [Ships]) and the Director, Surface Warfare (N86), of the two designs of the Littoral Combat Ship to deploy and operate the system mission module with attention to sea-state capability vis-à-vis safety, counterdetection of both the module and attending LCS in unprotected areas, system reliability of undersea arrays, manning requirements, logistics, and storage. If the capability provided by a deployable system such as the ADS is considered essential as an ASW mission package module, the size, manning requirements, organic shipboard data link and processing requirements, launching, handling, logistics, and stowage requirements and systems need to be reviewed, revalidated, and, if necessary, rebaselined. An assessment of other surface ships as delivery platforms should be included; and

• An assessment by the Deputy Chief of Naval Operations for Communication Networks (N6) and the Naval Network Warfare Command of the command, control, and communications aspects (specifically satellite communications) of the ADS.

Finding 4: DRS S&T efforts will bring incremental, and perhaps new, capabilities over the next 5 to 10 years or more. This is an area of much technical ferment in both the naval and scientific communities. Significant obstacles do remain concerning both the theory and the implementation of DRS systems. Additionally, there is no clear transition path from potentially useful S&T experiments into fielded naval warfare capability.

Certain technologies critical to future, broadly based DRS systems are not yet mature enough for routine operational use. Those with the highest potential payoff are automated DCLT algorithms for operator support for both fixed and mobile sensor fields, acoustic communications and undersea networks, autonomous operations and distributed autonomy for UUVs so that they can both close on a target and set up projected DCLT based on target course and speed, energy sources and their efficient use for both propulsion and electronics, and improved and possibly alternative sensors.

The committee also notes that technologies useful to DRS capabilities come from a large and diverse set of disciplines and vendors. While many are motivated by a range of naval needs, much of the expertise is not so specific; consequently, the technology base is fragmented. For a successful DRS S&T effort, there needs to be a continual awareness of relevant technologies. Therefore, the Department of the Navy needs to foster the growth of a DRS community that does the following: brings together outstanding technologists from relevant disciplines from both within and outside the department, including academia and industry; educates non-naval experts on naval needs; and encourages approaches to solving naval problems.

Recommendation 4: To ensure that DRS system capabilities continuously evolve and transition from potentially useful S&T experiments into fielded naval warfare capability, the Deputy Chief of Naval Operations for Integration of Capabilities and Resources (N8) and the Chief of Naval Research (CNR) should:

- Maintain and, if possible, increase 6.1 and 6.2 funding for DRS-related efforts; and
- Align relevant 6.2 and 6.3 funding to provide for technology insertion into the DRS program office advocated in Recommendation 1. This DRS program office should establish very close links between the S&T technologists and the Program Executive Office for Integrated Warfare Systems to provide both technology push and operational pull, and together with the CNR, it should coordinate efforts at the fundamental and applied level for the desired evolution of DRS systems.

Finding 5: Open architectures and an approach to interface standards that accommodates changes will significantly reduce the time and costs required for fielding DRS systems. Instead of fixing the problems retroactively, some forethought on these issues leads to cost savings as well as increased reliability. The following systems need to be made open and interoperable: planning, monitoring, control, sensor, communications, undersea situational awareness, network, and data representation. In addition, provision for operations with multiple levels of classification for system capabilities and data security are to be anticipated.

Virtually every large-scale-systems engineer has encountered the problems associated with closed software, incompatible operating systems, and data protocols. It is a bona fide and costly situation that leads to delays and degraded reliability and can be avoided by forward thinking. Inevitably there will be resistance from those who already have some stake in existing DRS systems, since these systems have been developed independently. Nevertheless, one of the tasks of the DRS program office advocated in Recommendation 1 should be the promulgation of open architectures and interoperability standards so that all will know that eventually these architectures and standards will be adopted to be part of the Department of the Navy's efforts.

Open and interoperable systems also pave the way toward more rapid fusion of underwater and above-water information: for example, surveillance and identification of ship tracks to eliminate potential submarine DCLT for a DRS system by aircraft or satellites, clarification of the common battlefield environment status, and the routing of important environmental data are all well-known attributes.

Recommendation 5: Since open architectures and interface standards are an important topic that becomes more difficult as different systems evolve, the ASN(RDA) should:

- Direct the CNR and the PEO(IWS) (or the new office) to
  —Establish standards for open architectures and interoperability for the planning of emerging DRS systems; and
  —Ensure that the S&T and R&D communities associated with networks, communications, command, and control start with simple systems and capabilities and develop them spirally rather than creating them under a large-system framework.
- Direct the PEO(IWS) (or the new office) to
  —Mandate open architectures for all DRS systems, including command and control;
  —Adopt Joint Unmanned Systems Common Control (JUSC2) as the framework for the monitoring and control of DRS systems; and
  —Adopt interoperable standards for acoustic communications and networking.

Finding 6: Efforts toward developing DRS systems for countermine warfare are important and, at the present time, are adequately implemented within the constraints of available technology and funding. Much remains to be done, however, to capture the Department of the Navy's future vision of countermine warfare.

As articulated by the Department of the Navy[7] and in the *FY07 U.S. Naval Mine Countermeasures (MCM) Plan*[8] by the Commander, Mine Warfare Command, the future vision of countermine warfare involves coordinated, cooperative activity by distributed, networked, and cooperative sensors and platforms to increase the speed and effectiveness of mine countermeasures. Previous reports of the Naval Studies Board[9] have suggested the use of DRS systems to detect and track mining operations in shallow-water areas eligible for amphibious landings. Existing programs for mine countermeasures (MCM) organic capabilities

[7]VADM Michael Bucchi, USN; and VADM Michael Mullen, USN. 2002. "Sea Power 21 Series, Part II: Sea Shield: Projecting Global Defensive Assurance," U.S. Naval Institute Proceedings, November, p. 58.

[8]Department of the Navy. 2006. *FY07 U.S. Naval Mine Countermeasures (MCM) Plan: Vision, Roadmap, and Program,* Washington, D.C., February.

[9]Naval Studies Board, National Research Council, 2001, *Naval Mine Warfare: Operational and Technical Challenges to Naval Forces,* National Academy Press, Washington, D.C.; Naval Studies Board, National Research Council, 1997, *Technology for the United States Navy and Marine Corps, 2000-2035: Becoming a 21st-Century Force, Volume 7: Undersea Warfare*, National Academy Press, Washington, D.C.; Naval Studies Board, National Research Council, 1994, *Mine Countermeasures Technology, Volume II: Task Force Report* (U), National Academy Press, Washington, D.C. (classified).

involving unmanned surface vehicles and UUVs can be harnessed to provide components and near-term DRS systems to speed up MCM operations. Also, ADS-type systems deployed for ASW in deeper-water areas may be able to detect and track submarine mining operations.

Recommendation 6: In its efforts to bring to fruition DRS systems for significant, near-term capabilities (see Recommendation 1), the Department of the Navy should examine the potential application of ASW DRS systems for countermine missions. Specifically,

- The Deputy Chief of Naval Operations for Integration of Capabilities and Resources (N8) should accelerate support toward achieving an organic MCM capability, and experiments and simulations should be performed to see how organic MCM components integrated into DRS configurations could speed up MCM operations; and
- The Program Executive Office for Littoral and Mine Warfare (PEO[LMW]) should review the 2001 Naval Studies Board report entitled *Naval Mine Warfare: Operational and Technical Challenges for Naval Forces* and the 2002 U.S. Fleet Forces Command report entitled *Mine Warfare—The Way Ahead: Long Range Planning and Concepts* to assess the implementation of the recommendations in these reports.

Appendixes

A

Committee and Staff Biographies

Arthur B. Baggeroer, Co-chair, is Ford Professor of Engineering and the Secretary of the Navy/Chief of Naval Operations Chair for Ocean Science in the Departments of Ocean and Electrical Engineering at the Massachusetts Institute of Technology (MIT). His areas of expertise include advanced signal-processing methods applied to sonar, ocean acoustics, and geophysics. Dr. Baggeroer has served as director of the MIT-Woods Hole joint program in oceanography and oceanographic engineering. During sabbatical leaves he has also served as a consultant to the Chief of Naval Research at the NATO Supreme Allied Commander Atlantic Center in La Spezia, Italy, and as a Green Scholar at the Scripps Institution of Oceanography. He is a fellow of the Institute of Electrical and Electronics Engineers and the Acoustical Society of America. Dr. Baggeroer is a member of the National Academy of Engineering and a former member of the National Research Council's (NRC's) Naval Studies Board.

Brig "Chip" Elliott, Co-chair, is principal engineer at BBN Technologies, where he has led the design and successful implementation of a number of secure, mission-critical networks based on novel Internet technology for the United States, Canada, and the United Kingdom. His areas of expertise include wireless Internet technology, mobile ad hoc networks, quality-of-service issues, and novel routing techniques. Mr. Elliott has served as a senior advisor on a number of national and commercial networks, including those for the Discoverer II, Space-Based Infrared System Low, and Celestri/Teledesic satellite constellations and Boeing's Connexion system. He has also served on numerous scientific boards and advisory committees, such as the Army Science Board and the Defense Science Board as well as several NRC committees. Mr. Elliott is a member of the NRC's Naval Studies Board.

James G. Bellingham is chief technologist at the Monterey Bay Aquarium Research Institute. His personal research interests revolve around the development and use of autonomous underwater vehicles (AUVs). Dr. Bellingham has spent considerable time at sea, leading more than 20 AUV expeditions in locations such as the Antarctic, North Atlantic, Mediterranean, South Pacific, and the Arctic. At present he is working on the development of distributed ocean observing systems, notably the Autonomous Ocean Sampling Network project. Dr. Bellingham is a founder of Bluefin Robotics Corporation, a leading manufacturer of AUVs for the military, commercial, and scientific markets.

E. Ann Berman is president of Tri-Space, Incorporated, a remote sensing and software-engineering company serving a broad range of environmental and security areas. Her areas of expertise include remote sensing, measurement intelligence, thermal and spectral analysis, and military space systems. She is currently working with the National Air and Space Intelligence Center (NASIC) on overhead non-imaging infrared processes for estimating diver visibility, light penetration, and other environmental information. Formerly, Dr. Berman served as Deputy Assistant Secretary of the Navy for C3I and Space, where she managed the Navy blue-green laser communications initiative. She has designed marketing strategies at Communications Satellite Corporation for store-and-forward distributed sensor arrays and a stereographic satellite payload for global terrain mapping. Dr. Berman has also served on numerous scientific boards and advisory committees, including the NRC's Committee on Environmental Information for Naval Use and the Committee on the Navy's Needs in Space for Providing Future Capabilities.

D. Richard Blidberg is director of the Autonomous Undersea Systems Institute. His areas of expertise include the development of technologies related to autonomous submersible vehicles and architectures for the intelligent guidance and control of multiple autonomous vehicles. Mr. Blidberg has served as manager of the seabed survey operations at Ocean Research Equipment, Inc.; served with the U.S. Coast Guard; and worked at the Woods Hole Oceanographic Institution. He has also served on numerous scientific and technical advisory groups. He is currently associate editor of the Institute of Electrical and Electronics Engineers *Journal of Ocean Engineering.*

Daniel R. Bowler retired from the U.S. Navy with the rank of rear admiral after serving more than 32 years as a surface warfare officer. He is vice president, Maritime Sensors and Electronics Systems, in the Washington Operations Office of Lockheed Martin Corporation, responsible for sensors and electronics for a wide variety of programs across the Services and the Department of Defense. Before joining Lockheed, Admiral Bowler served as director, Warfare Integration and Assessment, Office of the Deputy Chief of Naval Operations for Warfare, Requirements, and Programs. His background includes naval and joint undersea warfare missions, concepts of operations, and capabilities, including distributed remote sensing requirements.

David L. Bradley is a senior research scientist at the Applied Research Laboratory, Pennsylvania State University. His areas of expertise include environmental acoustics, in particular, conducting research in acoustic radiation, propagation, scattering, reflection, absorption, and natural/manmade noise analysis, including spatial and temporal fluctuations. Dr. Bradley is a former member of the NRC's Ocean Studies Board, chaired the Steering Committee for the Sixth Symposium on Tactical Oceanography, and was a member of the Committee on Potential Impacts of Ambient Noise in the Ocean on Marine Mammals.

Albert H. Konetzni, Jr., retired from the U.S. Navy with the rank of vice admiral after 38 years of service, including the position of Commander, Submarine Force, U.S. Pacific Fleet, and Deputy Commander and Chief of Staff, U.S. Atlantic Fleet. His background is in Department of the Navy and Department of Defense planning, programming, and budgeting activities, as well as naval operations involving joint and coalition forces capabilities.

William A. LaPlante is head of the Strategic Systems Department at the Applied Physics Laboratory at the Johns Hopkins University. His areas of expertise include technologies associated with antisubmarine warfare, including passive and active acoustics in the nuclear-powered ballistic missile submarine security program. Dr. LaPlante served on the Chief of Naval Operations' Task Force Antisubmarine Warfare.

Terry P. Lewis is a senior systems engineer with the Raytheon Company. His areas of expertise include command, control, communications, and information systems; digitized battlespace systems; communications and transmission security in military tactical systems; wireless network security; and network management authentication techniques for robust security architecture. Mr. Lewis has developed antitampering technologies to prevent or reduce the ability of potential aggressors to reverse-engineer critical U.S. communications technologies. He is a Raytheon fellow and received the Most Promising Engineer of the Year award conferred at the 2002 Black Engineer of the Year Award Conference.

Thomas V. McNamara is director of the Tactical Systems Programs Office at the Charles Stark Draper Laboratory, where his primary focus is the development of manned and unmanned systems and weapons systems. These two areas include the development of technologies and systems to address the challenges of littoral, command-and-control, precision engagement, and missile defense systems. Mr. McNamara's areas of expertise include guidance, navigation, and control; intelligent autonomy; precision weapons delivery; microelectromechanical sensors; dismounted-soldier systems; mission planning; and systems integration for naval submersible and aircraft platforms. He is a member of the NRC's Naval Studies Board.

L. David Montague, an independent consultant, is retired president of the Missile Systems Division at Lockheed Martin Missiles and Space and a former officer of Lockheed Corporation. His areas of expertise include the design, development, and program management of military weapons and their related systems, as well as complex systems, engineering and systems integration of ballistic missiles, cruise missiles, and unmanned aerial vehicles. Mr. Montague has served on numerous scientific boards and advisory committees, including the Navy Strategic Systems Steering Task Group and task forces for both the U.S. Army and the Defense Science Board. He currently serves as a member of the director's senior advisory board of the Los Alamos National Laboratory. He is a fellow of the American Institute of Aeronautics and Astronautics (AIAA) and a previous recipient of the AIAA's Missile Systems Award. Mr. Montague is also a member of the National Academy of Engineering and the NRC's Naval Studies Board.

Douglas R. Mook is president of the Aptec Group. His areas of expertise include acoustic processing and sensor fusion, advanced acoustics communications, digital battlefield programs for communications and sensors, and unattended ground sensors. He has served as chief operating officer and vice president of engineering at TechOnLine and managed the Antisubmarine Warfare Division, advanced systems, advanced technology, and signal-processing organizations within Lockheed Martin. He has also served as a member of the NRC's Committee on Network-Centric Naval Forces, as well as a member of the Navy's Fleet Ballistic Missile Submarine Review Panel and the Army's Digital Battlefield Definition Panel.

John E. Rhodes retired from the U.S. Marine Corps with the rank of lieutenant general, having served as commanding general of the U.S. Marine Corps Combat Development Command (MCCDC). While at MCCDC, General Rhodes led the Marine Corps in its development of warfighting concepts and in the integration of all aspects of doctrine, organization, training and education, equipment, and support and facilities enabling the Marine Corps to field combat-ready forces. This responsibility entailed, among other things, careful assessments of current and future operating environments and the continuous adaptation of the Corps's training infrastructure and resources in order to ensure that the integrated capabilities were continuously developed for the Unified Commanders in Chief.

James Ward is leader of the Advanced Sensor Techniques Group at the MIT Lincoln Laboratory, where he has worked since 1990. His areas of technical expertise include signal processing for radar, sonar, and communications systems; adaptive array and space-time adaptive processing; detection and estimation theory; and systems analysis. Dr. Ward received his B.E.E. from the University of Dayton (Ohio) in 1985 and his M.S.E.E. and Ph.D. from Ohio State University in 1987 and 1990, respectively. In 2001 he was the recipient of the MIT Lincoln Laboratory Technical Excellence Award, and in 2003 he received the IEEE Aerospace and Electronic Systems Society's Fred Nathanson Young Engineer Award for contributions to adaptive radar and sonar signal processing. Dr. Ward is a fellow of the IEEE.

Dana R. Yoerger is an associate scientist of the Deep Submergence Laboratory, Department of Applied Physics and Engineering, at the Woods Hole Oceanographic Institution. His areas of expertise include underwater vehicles and manipulators, telerobotic systems designed for seafloor survey, autonomous vehicles used for the long-term monitoring of the deep ocean, vehicle and tether dynamics, and the application of modern nonlinear and adaptive control techniques to underwater vehicle operation. Dr. Yoerger served on the NRC's Committee on Undersea Vehicles and National Needs.

Staff

Charles F. Draper is director of the NRC's Naval Studies Board. Before joining the NRC in 1997, Dr. Draper was the lead mechanical engineer at S.T. Research Corporation, where he provided technical and program management support for satellite Earth station and small satellite design. He received his Ph.D. in mechanical engineering from Vanderbilt University in 1995; his doctoral research was conducted at the Naval Research Laboratory (NRL), where he used an atomic-force microscope to measure the nanomechanical properties of thin-film materials. In parallel with his graduate student duties, he was a mechanical engineer with Geo-Centers, Incorporated, working on-site at NRL on the development of an underwater X-ray backscattering tomography system for nondestructive evaluation of U.S. Navy sonar domes on surface ships.

Arul Mozhi is senior program officer at the NRC's Naval Studies Board; he also served as senior program officer at the NRC's Board on Manufacturing and Engineering Design and National Materials Advisory Board. Prior to joining the NRC in 1999, Dr. Mozhi was senior scientist and program manager at UTRON, Inc., a high-tech company in the Washington, D.C., area, working on pulsed electrical and chemical energy technologies applied to materials processing. From 1989 to 1996, Dr. Mozhi was a senior engineer and task leader at Roy F. Weston, Inc., a leading environmental consulting company working on long-term nuclear materials behavior and systems engineering related to nuclear waste transport, storage, and disposal in support of the U.S. Department of Energy. Before 1989 he was a materials scientist at Marko Materials, Inc., a high-tech firm in the Boston area, working on rapidly solidified materials. He received his M.S. and Ph.D. degrees (the latter in 1986) in materials engineering from Ohio State University and then served as a postdoctoral research associate there. He received his B.S. in metallurgical engineering from the Indian Institute of Technology in 1982.

B

Summary of Committee Meeting Agendas

The Committee on Distributed Remote Sensing for Naval Undersea Warfare first convened in June 2005 and held additional meetings and site visits over a period of 7 months:

- *June 23-24, 2005, in Washington, D.C.* Organizational meeting: Office of the Chief of Naval Operations, Fleet ASW Command, Program Executive Office for Integrated Warfare Systems, and Program Executive Officer for Littoral and Mine Warfare briefings on requirements for distributed remote sensing for naval undersea warfare.
- *July 14-15, 2005, in Washington, D.C.* Space and Naval Warfare Systems Command, Office of the Chief of Naval Operations, and Naval Air Systems Command briefings on requirements for distributed remote sensing for naval undersea warfare; and Defense Advanced Research Projects Agency and Office of Naval Research briefings on distributed remote sensing programs.
- *August 17-18, 2005, in Woods Hole, Massachusetts.* Naval Undersea Warfare Center, Navy Warfare Development Command, Lockheed Martin Corporation, Applied Research Laboratory/University of Texas at Austin, Applied Physics Laboratory/Johns Hopkins University, Dynamics Technology, Inc., Applied Research Laboratory/Pennsylvania State University, Electric Boat Corporation, BAE Systems, Raytheon Company, and Boeing Integrated Defense Systems briefings on distributed remote sensing programs.
- *September 22-23, 2005, in Washington, D.C.* Fleet Anti-Submarine Command, Space and Naval Warfare Systems Command, National Science Foundation, Charles Stark Draper Laboratory, National Defense Industrial Association,

and Northrop Grumman Corporation briefings on distributed remote sensing studies and programs.

- *October 27-28, 2005, in Washington, D.C.* Naval Surface Warfare Center/ Coastal Systems Station, Program Executive Office for Littoral and Mine Warfare, Naval Meteorology and Oceanography Command, Defense Advanced Research Projects Agency, Lincoln Laboratory/Massachusetts Institute of Technology, Exponent, and Monterey Bay Aquarium Research Institute briefings on distributed remote sensing programs.
- *December 8-9, 2005, in Washington, D.C.* Office of Naval Research; Office of Naval Intelligence; Program Executive Office for Integrated Warfare Systems, Underwater Systems; Department of Homeland Security; Marine Physical Laboratory/University of California at San Diego; Massachusetts Institute of Technology; AETC, Inc.; and Sverdrup, Inc., briefings on distributed remote sensing requirements and programs.
- *January 24-27, 2006, in Irvine, California.* Committee deliberations and report drafting.

C

Acronyms and Abbreviations

ACINT	acoustic intelligence
ACOMMS	acoustic communications
ADS	Advanced Deployable System
APB/ARCI	Advanced Processor Build/Acoustic Rapid COTS Insertion
ASN(RDA)	Assistant Secretary of the Navy for Research, Development, and Acquisition
ASW	antisubmarine warfare
CFFC	Commander, Fleet Forces Command
CNO	Chief of Naval Operations
CNR	Chief of Naval Research
CONOPS	concept of operations
COTS	commercial off-the-shelf
DASN(IWS)	Deputy Assistant Secretary of the Navy for Integrated Warfare Systems
DASN (Ships)	Deputy Assistant Secretary of the Navy for Ships
DCLT	detection, classification, localization, and tracking
DODAF	Department of Defense Architecture Framework
DRS	distributed remote sensing
EER	Extended Echo Ranging
FDS	Fixed Distributed System
ISR	intelligence, surveillance, and reconnaissance
JUSC2	Joint Unmanned Systems Common Control
LCS	littoral combat ship
MCM	mine countermeasures
MRUUVS	Mission Reconfigurable UUV System

N6	Deputy Chief of Naval Operations for Communication Networks
N8	Deputy Chief of Naval Operations for Integration of Capabilities and Resources
N86	Director, Surface Warfare Division (reporting to N8)
NRC	National Research Council
NWDC	Navy Warfare Development Command
ONR	Office of Naval Research
PEO(IWS)	Program Executive Office for Integrated Warfare Systems
PEO(LMW)	Program Executive Office for Littoral and Mine Warfare
R&D	research and development
RF	radio frequency
SNR	signal-to-noise ratio
SSGN	nuclear-powered, guided-missile submarine
SSN	nuclear-powered submarine
S&T	science and technology
SURTASS	Surveillance Towed Array Sensor System
USW	undersea warfare
UUV	unmanned undersea vehicle